Probability and Intuition

Sujith Vijay

ISBN-13: 979-8685572875

Cover image: *The 1821 Derby at Epsom* by Theodore Gericault

Dedicated to the loving memory of my brother Renjith Vijay (1985 - 2015)

CONTENTS

Chapter 1: Correlation and Causation

In 2005, the Kansas School Board of Education received an open letter from a young science graduate named Bobby Henderson. The letter urged the authorities to consider teaching, in addition to the Christian creationist view recommended by the board as an alternative to Darwin's theory of evolution, other theories of intelligent design as well. Specifically, Henderson put forth the tenets of a new religion called *Pastafarianism*, which expounded that the universe and everything in it was created by an omnipotent entity known as the Flying Spaghetti Monster. The usual prescriptions and proscriptions that accompany every other religion were present in Pastafarianism as well – for example, it was forbidden to teach the beliefs of the religion without wearing His chosen outfit, namely full pirate regalia.

Among the many memorable asides in Henderson's hilarious letter was the following observation:

"*You may be interested to know that global warming, earthquakes, hurricanes, and other natural disasters are a direct effect of the shrinking numbers of pirates since the 1800s. For your interest, I have included a graph of the approximate number of pirates versus the average global temperature over the last 200 years. As you can see, there is a statistically significant inverse relationship between pirates and global temperature.*"

While the graph is omitted here, the reader can easily see the (intentional) flaw in this sort of reasoning. Just because there is a direct or inverse relation between two variables, it does not mean that the first one caused the second. It could well be that the two events have no causal relation whatsoever, or the second one actually caused the first, or the two events have a common cause (in the pirates example, the second order effects of the industrial revolution). The usual mantra that encapsulates this idea, drilled in a first year college level course in statistics is: *"Correlation does not imply causation."*

However, causation (or causality, to use the more common term) is notoriously difficult to establish. It is a metaphysical conundrum that has vexed great philosophers like David Hume and Immanuel Kant, and remains a feral animal to this day. Hume believed that causal inference was contingent on experience, and any causality established based on a sequence of observations is, at best, predicated on the hope that the future will continue to resemble the past. Causality, according to Hume, is imputed by the observer based on the constant conjunction of events, and is really a leap of faith. Kant, on the other hand, postulated that causality exists independent of experience, as what he calls a *synthetic a priori judgement*. Causality, according to Kant, is an additional feature of the same cognitive apparatus that allows an observer to infer which of the two events occurred before the other. Centuries later, the jury is still out on whether Kant resolved Hume's objection, or even just what Kant intended in the first place.

A contemporary philosopher and computer scientist who has done highly influential work on causality is Judea Pearl, director of the Cognitive Systems Laboratory at the University of California, Los Angeles. Pearl's working definition of causality is that it would predict the consequences of intervention in a system.

A causes B if and only if we can modulate B by altering A. Ideally, A and B are parametrized by continuous variables, and one can check if changes in the purported cause lead to commensurate changes in the observed effect. Thus not only does gravity *cause* a ripe apple to fall to the earth's surface, we can even predict its rate of change of velocity in terms of the mass and radius of the earth. (An experimenter may not be able to do much about the mass of the earth, but the effective radius does depend on latitude.) The probabilistic analogues of such questions are far more intricate, but Pearl, building upon prior work by Peter Spirtes, Clark Glymour and Richard Scheines, has developed a *causal calculus* using combinatorial objects called *directed acyclic graphs* to deal with them. This is the state of the art when it comes to answering instances of the question, *"Is this just correlation or can it be upgraded to causation?"*

It is now accepted scientific wisdom that smoking causes lung cancer, but this was a hard-won victory. It was bad enough that the tobacco lobby fought tooth and nail against the advancement of science, by funding study after study to reshape consensus. But they also found an unlikely ally in Sir Ronald Fisher, arguably the greatest statistician in history and an unapologetic pipe smoker, who steadfastly opposed the conclusions of the landmark study of Richard Doll and Austin Hill that made a highly persuasive argument for the causal connection between smoking and lung cancer. In a letter to the British Medical Journal in 1957, Fisher conceded that there was a *prima facie* case for further investigation, but refused to accept the question as settled. He had high standards for causation, and apparently the study had not met them. As Fisher died a few years later, it is unclear if he would have changed his mind as the mountain of evidence grew in the subsequent decades. In any case, Pearl's causal calculus now provides a framework to address these and other questions. The moral of the story is that one should not be too disdainful or dismissive of correlation. As the American statistician Edward Tufte observed, *"Correlation is not causation but it sure is a hint."*

Chapter 2: Regression to the Mean

In 1885, British statistician and anthropologist Sir Francis Galton published the results of a pioneering study that compared the heights of 898 adult children with that of their parents. The purpose of the study was to investigate the heritability of height. Indeed, taller parents did have taller children on average, and shorter parents had shorter children on average. Thus height was found to be a heritable trait within the confidence level offered by the sample size. But the study is remembered today for the surprising auxiliary finding that the children of taller parents were typically shorter than their parents, while the children of shorter parents were typically taller than their parents. In other words, the data did not support the thesis that the progeny of outliers would continue to be outliers. If anything, it indicated just the opposite. Galton called this phenomenon regression to mediocrity, and these days it is called *regression to the mean*.

In some sense, Galton's finding is not surprising at all, as a lower bar (of parental height) is easier to cross than a higher bar. However, given that height is heritable, the lower bar could well be as difficult for the child of shorter parents as the higher bar is for the child of taller parents. There are two opposite effects at play here, and only a statistical study can bring out their respective strengths. (For simplicity, we are ignoring situations where one parent is relatively tall and the other relatively short, though of course Galton's data set had plenty of such cases.) It is worthwhile to mention that regression to the mean should not be confused with the *gambler's fallacy*, which erroneously presumes that a fair coin is more likely to show tails if, say, the previous three tosses showed heads.

The basic idea behind regression to the mean is that records are set when skill (or more generally, fitness) accompanies luck, and while skill is relatively stable, luck varies all over the place. A memorable instance happened in 1934, when Italian striker Raimundo Orsi scored a late equalizer through a beautiful curling shot in the World Cup Football final match against Czechoslovakia. (Italy ended up winning 2-1 in extra time.) Since television coverage was in its infancy during those days, journalists requested him after the match to demonstrate it one more time for the cameras. Orsi tried many times in vain to recreate his brilliant shot, and finally gave up. He was lucky when it mattered with eleven people out to stop him, but now the Gods were schooling him in gratitude.

A common example of regression to the mean is when governments take credit for policy interventions after a natural or man-made disaster. Extreme bad luck is a prerequisite for most disasters, like five different things going wrong at the same time, so a government that only wakes up after the event and puts some mitigating safety measures in place may mistakenly believe that these steps are preventing the recurrence of a once in a century event. Of course, the simpler explanation is the fact that such events typically occur only once in a century. This is not to say that precautionary measures should not be taken after a disaster – they most certainly should be. But it would be even better if they are taken in anticipation, not after the horses have bolted.

Chapter 3: The Battle of Inferences

Mathematicians usually take pride in the certainties offered by their subject. Indeed, apart from occasional priority disputes, like the one between Newton and Leibniz over the discovery of calculus, there have hardly been any real controversies in the history of mathematics. The only noteworthy ideological battle was between the *classical* school of David Hilbert and the *intuitionist* school of L.E.J. Brouwer in the 1920s regarding the reliability of *the law of excluded middle* as a rule of logical inference, which was soon forgotten after more serious foundational issues threatening both belligerents were raised by Kurt Gödel. In recent years, however, a new ideological schism has arisen in the statistics and machine learning community – the battle between *frequentist inference* and *Bayesian inference.*

The differences between the camps are subtle but significant. Frequentists, as the name suggests, consider probability as a measure of frequencies in the long run. Bayesians consider probability as a subjective measure of belief, to be updated as new data comes in. They begin with an assumed *prior distribution* and update the posterior probability using the Bayes' rule of conditional probability, namely:

$$P(A|B) = \frac{P(B|A)P(A)}{P(B)}$$

As seen above, Bayes' rule links *P(A|B)*, the probability of the event A occurring given that B has occurred, with *P(B|A)*, the probability of the event B occurring given that A has occurred. The goal of Bayesian inference is to compute the conditional probability distribution of the unknown parameter given the observed data, in terms of the conditional probability of the observed data given that the unknown parameter equals p.

Let us take a concrete example. Suppose we have a coin with an unknown probability p of showing heads on tossing. We toss the coin 100 times and observe 52 heads. It would be tempting to conclude that the coin is actually fair, and the small discrepancy is just random fluctuation. However, the value of p that best explains the observations is not 0.5, but $52/100=0.52$, or k/n in general if n is the number of tosses and k is the number of heads. A frequentist would conclude that the best estimate for p is 0.52, and will come up with an appropriate *confidence interval* around 0.52, such that one can be, say, 95% confident that the true value lies in this interval. In other words, if the experiment is performed a sufficiently large number of times, the confidence interval will contain the true value at least 95% of the time.

Bayesians need a prior distribution to start, just like the well-known Newton-Raphson root finding method needs an initial guess. Let us assume that the unknown probability p is uniformly distributed on the interval (0,1). For each value of p, there is a posterior distribution obtained using Bayes' rule. Integrating over the interval (0,1) to average over all possible values of p, the posterior expectation is obtained to be 53/102, or $(k+1)/(n+2)$ in general. Intuitively, this is as if we first run a trial phase to make sure both heads and tails show up, and incorporate the results of the last two tosses from the trial phase (one of which will be head and the other tail) into our experiment. This correction to the frequency estimate is called *Laplace's rule of succession*, named after mathematician Marquis de Laplace's famous estimate of the probability that the sun will not rise tomorrow, given that it has done so for every day in the past 5000 years. Laplace reasoned that the probability was tiny, but not zero.

Rather than confidence intervals, Bayesians tend to prefer *credible intervals* around their point estimate, inside which the unknown parameter falls with, say, probability 0.95. Since Bayesian inference already returns a posterior distribution, this interval can be computed quite naturally, excluding the left and right tails. Of course, the credible interval depends on the prior distribution.

So who is right? It doesn't really matter. This is not about truth, but about imperfect estimates we have to make about an unknown and possibly unknowable truth, with limited information. Frequentists treat parameters as unknown constants, whereas Bayesians treat parameters as random variables. The differences between the two approaches are ultimately a matter of interpretation, and their intuitive appeal depends heavily on context.

The situation is analogous to the well-known conundrum in probability theory known as *Bertrand's paradox*. The innocuous question of determining the probability that a randomly chosen chord of a circle is longer than the side of the equilateral triangle inscribed in the circle turns out to be ill-defined. There are three perfectly reasonable approaches that yield the answers 1/2, 1/3 and 1/4. There is nothing wrong with any of these answers, as they are based on three different ways of choosing the random chord. The question is well-defined only if the phrase *"randomly chosen chord"* is clarified in terms of operational details, i.e., just how is the chord chosen. The battle between frequentists and Bayesians is similar in spirit, but there are practitioners who defend their positions with admirable zeal, providing much needed entertainment to a rather demanding subject.

Chapter 4: Skewness and Kurtosis

Many interesting probability distributions, including the ubiquitous normal distribution, are symmetric about the mean. But there are many other examples, like the exponential and Poisson distributions, that are not symmetric. A quantitative measure of the asymmetry of the right and left halves of a probability distribution is *skewness*, defined as follows:

Recall that the Taylor's theorem in elementary calculus allows us to compute the value of an infinitely differentiable function at any point, under some mildly restrictive assumptions, using the value of the function and all its derivatives at a single point. In probability theory, the so-called moments play a similar role to derivatives. Let X be a continuous random variable with probability density function f(x). Then the average value or *expectation* of the random variable, denoted E(X), is computed as follows:

$$E(X) = \int_{-\infty}^{\infty} x\, f(x)\, dx$$

The expectation is also called the *mean* of X, or the *first moment* of X. In the continuous case, it is a probability weighted integral, and in the discrete case, it is a probability-weighted sum. Let Y=X-E(X). Since the expectation of the difference of two random variables equals the difference of the individual expectations, it follows that E(Y)=0. We say that the *first central moment* of every random variable equals zero.

The expectation of Y^2, or the *second central moment* of X is called the *variance* of X. The square root of variance is called standard deviation, denoted by σ. Let Z = Y/σ. Then E(Z^2)=1. We say that the *second standardized moment* of every random variable equals 1.

Skewness is the expectation of Z^3, or the *third standardized moment* of X. We say that X is positively skewed (or right-skewed) if $E(Z^3) > 0$, and negatively skewed (or left-skewed) if $E(Z^3) < 0$. The skewness of a symmetric distribution, such as the normal distribution, is zero. It is usually the case that the mean of a random variable with a right-skewed distribution is greater than the median, and the mean of a random variable with a left-skewed distribution is less than the median. However, exceptions to this rule do exist, and the interested reader can refer to the comprehensive article titled *Mean, Median and Skew: Correcting a Textbook Rule* by Paul von Hippel, published in 2005 in the *Journal of Statistics Education* (Volume 13, Number 2) for details.

The importance of skewness is encapsulated in the following advice to investors, speculators, gamblers and various mixtures thereof, attributed to legendary hedge fund manager George Soros: *"It doesn't matter how often you are right or wrong – it only matters how much you make when you are right, versus how much you lose when you are wrong."* Even if the median earnings from a sequence of investments or bets is positive, one could go bust if the average earnings are negative. This is often how insurance companies collapse in the aftermath of natural disasters. They collect healthy premiums for decades, and then end up making huge payouts to a large number of clients in a matter of hours. Homeowner insurers in the state of Louisiana had made a cumulative profit of $1 billion from 1980 to 2005. On 29 August 2005, the day hurricane Katrina hit the city of

New Orleans, they lost $8 billion. Skew comes at you fast, and wreaks havoc all along its path. This is why the mitigation and management of risk is such a vital life skill, although most people realize its importance the hard way.

The expectation of Z^4, or the *fourth standardized moment* of X is called *kurtosis*. The kurtosis of the (univariate) normal distribution equals 3. Distributions with kurtosis greater than 3 are called *leptokurtic* and those with kurtosis less than 3 are called *platykurtic*.

Leptokurtic distributions are rediscovered every ten years or so by engineers working in mathematical finance, as they scratch their heads trying to figure out how they lost billions of dollars because of an event that was supposed to happen only once in a million years. If the underlying random variable is normally distributed and there is one event per day, then one typically expects the random variable to deviate from the mean by more than 6 standard deviations (a 6-sigma event) only once every 1.4 million years. Yet, David Viniar, CFO of Goldman Sachs, ruefully noted in August 2007: *"We were seeing things that were 25 standard deviation moves, several days in row."* Subsequent events, including the bankruptcy of Lehman Brothers brought home the truth that financial time series data are seldom amenable to simplistic normal distribution assumptions, and are heavily leptokurtic. Informally, the underlying distributions are *fat-tailed*, and rare events happen much more frequently than normal distribution assumptions would suggest.

Chapter 5: Overfitting

Among the mysteries in the history of scientific progress is the fact that the geocentric model of planetary motion with the sun and planets orbiting a stationary earth, advanced by Claudius Ptolemy in the 2nd century A.D. in his magnum opus *Almagest*, survived as long as it did. Visionaries like Aristarchus of Samos had proposed the heliocentric model as early as the 3rd century B.C., yet it was only in 1543, with the publication of *De Revolutionibus Orbium Coelestium* by Nicolaus Copernicus shortly before his death, that scholars began to publicly wonder if Ptolemy had got it wrong after all. The entire history of pre-Copernican astronomy is marred by *ad hoc* constructions of epicycles and deferents (essentially, circles rolling on circles) that became necessary to fit the geocentric model to observed data. Though it would mostly be an anachronistic objection, one can say that Occam's razor – the cautionary admonition by 13th century English friar William of Occam not to multiply entities without necessity – was thrown to the winds. Perhaps the greatest strength of Ptolemy's model was not its compatibility with observations but its unintentional affirmation of biblical scripture, such as the sun standing still on Joshua's command.

In retrospect, sufficiently many epicycles would indeed have approximated the elliptical orbits of planets to arbitrary precision. This is analogous to the theory of Fourier series in modern mathematics, which allows any periodic function (for example, a square wave) to be expressed as an infinite sum of sine and cosine functions of various frequencies. However, to borrow a beautiful quote from Richard Hamming,

"The purpose of computing is insight, not numbers." No zoological park of epicycles and deferents can stand in for the illumination provided by Kepler's elegant laws of planetary motion.

Overfitting existing data to a widely flexible model with a large number of parameters continues to be a vexing problem in the 21st century as well, most notably in the context of machine learning. The trouble is that in nearly all applications, the data contains signal as well as noise, and the task is to model the signal and disregard the noise. If there are too many parameters, one ends up modelling the noise in the past data as well. Good luck seeing the noise again in future data in identical shape and form just so the predictions come out all right!

So how does one prevent overfitting? A standard tool used in machine learning is *regularization*, which is just a quantitative version of Occam's razor, imposing a penalty on more complex models. Complexity may depend on some suitably defined norm (for example, the integral of the absolute value of the function). In the case of polynomials, a uniform upper bound on the absolute value of the coefficients may be imposed. The number of parameters are also kept under control using heuristics like the Akaike Information Criterion and Bayesian Information criterion, defined below.

The Akaike Information Criterion (AIC) attempts to minimize the difference between the number of parameters and the natural logarithm of the likelihood function (the probability of obtaining the observed sample given the parameters). Formally, $AIC = 2k - 2 \ln \hat{L}$ where k is the number of parameters and $\hat{L}$ is the maximum value of the likelihood function. Thus a new parameter is added to the model only if there is a commensurate increase in the logarithm of the likelihood that justifies its inclusion.

The Bayesian Information Criterion (BIC) has a very similar objective function, but it also accounts for the size.of the sample To be precise, $BIC = k \ln n - 2 \ln \hat{L}$ where k and $\hat{L}$ are as before and n is the sample size. The Bayesian Information Criterion is only used when the number of data points is much larger than the number of parameters. Note that both these criteria are used to select the best model from an available collection with no performance guarantee on the winner. If all input models are bad, the winner will naturally turn out to be bad, and neither criterion will give any warning in this regard.

Chapter 6: Precision and Recall

The central problem in jurisprudence and law enforcement is the tradeoff between the probability of convicting the innocent versus that of acquitting the guilty. Most cultures gravitate towards leniency, with variations on a sentiment expressed by the eminent British jurist William Blackstone: *"It is better that ten guilty persons escape than that one innocent suffer."* The 10:1 ratio championed by Blackstone was, of course, just a rule of thumb, and there is a similar quote in a letter by Benjamin Franklin that uses 100:1 instead. Authoritarian politicians have sometimes taken the opposite stand – a famous example is the legendary German Chancellor Otto von Bismarck, who advocated a 1:10 ratio. Mercifully, such instances are rather rare in modern democratic societies.

In the language of statistics, misclassifications such as these are usually designated Type I and Type II errors. A Type I error is a *false positive,* analogous to convicting an innocent person, or detecting an illness that is actually absent. A Type II error is a *false negative*, analogous to acquitting a guilty person, or failing to detect an illness that is actually present. Though we would ideally like to avoid both errors, an attempt to control one type of error often results in an increase in the frequency of the other type. What we do then depends on context. For example, in a legal setting, Type I errors are usually considered more serious, while in a medical setting, Type II errors are considered more serious.

Additional subtleties arise, for example, when we screen an entire population for a relatively uncommon illness. Consider the case of screening a population for COVID-19 as the cases are slowly beginning to rise. Suppose the prevalence is known to be around 4000 cases per million, based on (expensive but error-free) gold standard testing of a sample of sufficient size. Now suppose additionally that we use a less reliable test for screening larger groups, which has a 90% true positive rate (also called sensitivity or *recall*) and a 95% true negative rate (also called specificity). What should we expect if we test 10000 people?

In a typical scenario, we expect 40 of them to actually have the disease, 36 of whom will be detected correctly by the test. But out of the 9960 people who do not have COVID-19, as many as 498 will be misdiagnosed as having the disease. Thus we will end up with 534 people flagged as positive, out of which only 36 really have the disease. On the bright side, all but four of the 9466 people flagged as not having the disease will actually be disease-free. Thus the test has a very impressive negative predictive value of 99.96%, but an abysmally low positive predictive value of 6.74%. This is why routine screening of asymptomatic patients is not considered a good idea in the context of COVID-19 and many other diseases, especially in the initial stages when the prevalence is low.

The positive predictive value is also called *precision*. As shown in the above example, even the combination of high recall and high specificity is insufficient to guarantee high precision. The erroneous assumption of high precision in such circumstances is called the *base rate fallacy*. It should be noted, however, that a high negative predictive value can be quite useful in many contexts, especially when the isolation of large groups for monitoring and follow-up is impractical.

Chapter 7: The Kelly Criterion

Suppose we are asked to make bets on the outcome of twenty successive tosses of a biased coin, where the probability of the coin showing heads on each toss equals 0.6. (Never mind why someone would offer such a bet, although this is certainly something that should be asked in real life.) Obviously it makes sense to always bet on heads, but if the initial capital is $1000, how much money should be bet on each toss to maximize the expected winnings? Clearly, since we are more likely to win than lose, we can afford to be somewhat brave. But there is a fine line between brave and foolhardy, and that is what we are trying to figure out.

Suppose we bet half of our current capital on each bet. So in a typical scenario, our capital gets multiplied by

1.5 twelve times, and gets halved eight times. Since $(1.5)^{12} (0.5)^8 = 0.5068$, we surprisingly end up losing nearly half our capital in spite of the odds being in our favour. Clearly, betting 50% of our bankroll is way too aggressive.

What if we only bet 5% of our current capital on each toss? Then our final earnings get magnified by $(1.05)^{12} (0.95)^8 = 1.1914$ in a typical scenario. In other words, we make a return of 19% on our original investment. Not bad, but can we do better?

Let us work in a general setting. Suppose we have N bets and our probability of winning each bet is p. We will assume that $p > 0.5$, as we should not be betting at all if this is not the case. (Note that this only applies when the potential gain and potential loss on each bet are equal; otherwise betting on the less likely outcome could very well yield a positive expectation.) Let x be the fraction of our capital that we bet each time. Then the typical value W of our final wealth, starting from an initial capital W_0 is

$$W = W_0(1 + x)^{Np}(1 - x)^{N(1-p)}$$

Taking logarithms, this can be written as,

$$\frac{\log\left(\frac{W}{W_0}\right)}{N} = p \, \log(1 + x) + (1 - p) \log(1 - x)$$

Using elementary calculus, it can be shown that W is maximized when x = 2p-1. For our example, since p=0.6, the best strategy is to bet 20% of the capital on each toss, yielding a magnification of $(1.2)^{12} (0.8)^8 = 1.4959$ or a return on investment of nearly 50%. However, note that this is only the typical or *most probable* growth rate, and not the expected or *average* growth rate. If we end up with eleven wins and nine losses, which is not at all unlikely, we will actually lose 0.3%.

The optimization of the expected growth rate is a lot more complicated, and involves maximizing the expectation of the *logarithm* of the wealth rather than the wealth itself. But it turns out that the expected growth rate is also maximized when x=2p-1. This was first proved by John Kelly, an engineer at Bell Labs in 1956, and this optimal betting scheme is named the *Kelly criterion* in his honour.

In practice, even informed participants often place their bets well below the Kelly limit, as the additional margin of safety is well worth the disappointment of foregoing some of the potential upside. After all, in most real life scenarios, you only *think* you know the probability of winning. Moreover, expectation calculations converge only in the limit, as dictated by the *law of large numbers*, and the sequence of bets could end before convergence kicks in.

Related to the Kelly criterion is the idea of *variance drain*. This is the phenomenon where the rate of return of an investment scheme with variable returns falls below the rate of return of a fixed investment scheme, even if the arithmetic mean of the individual returns may be higher for the scheme with variable returns. Thus it is better to invest in a three-year fixed income scheme that yields a steady annual return of 8% rather than a stock that returns 21% in the first year, 3% in the second year and 1% in the third year. Although the arithmetic mean of the returns from the stock is 8.33%, the geometric mean of the returns is only 7.97%. On the other hand, the arithmetic mean and geometric mean of the returns are both 8% for the fixed income scheme. Ultimately, the geometric mean of the returns represents the true rate of growth of the investment, and this is also why one should never bet the entire capital even if the odds appear heavily favourable. The arithmetic mean of hundred and zero is fifty, while the geometric mean of hundred and zero is zero!

Chapter 8: The St. Petersburg Paradox

Suppose you walk into a casino with $100, and bet $1 on red on the roulette wheel. Since 18 of the 37 slots on the roulette wheel are red, you will win $1 with probability 18/37, and lose $1 with probability 19/37. You quit if you win; if you lose, you bet $2 on red. Again, with probability 18/37 you will win ($2 this time) and with probability 19/37 you will lose. If you lose, you bet $4 on red, and if you lose you bet $8 next time and so on. Eventually, red will show up, at which point you will recover the losses of all your previous bets and end up with a gain of $1. Sounds like a great plan, doesn't it?

This is called a *martingale betting strategy*, and the flaw is obvious. You only have $100, so if you lose six times in a row, you cannot place the seventh bet and must quit the game. While it is true that the probability of this happening is only $(19/37)^6 = 0.0183$ or about 1.8%, you stand to lose \$63 if this rare event does occur. Do you really want to participate in a scheme where you have a 98.2% chance of winning \$1 and a 1.8% chance of losing \$63? The expected value of the winnings is clearly negative, and the worst-case loss is 63 times the maximum possible amount you can win.

This last point is worth emphasizing. Even lotteries have a negative expectation from the point of view of the buyer, yet they are quite popular. Although the expectation is negative, the best-case gain is often millions of times the worst-case loss. A branch of behavioural economics called *prospect theory*, developed by Daniel Kahneman (winner of the Nobel Memorial Prize in Economics in 2002) and Amos Tversky, discusses this asymmetry of gain and loss as well as other deviations from perfectly rational behaviour observed in human decision making. In fact, the same idea can be seen in the classic work *The Theory of Games and Economic Behavior* by John von Neumann and Oskar Morgenstern, published in 1944.

Anyhow, the real catch is that nobody is infinitely rich, and given any fixed capital, there is a positive probability of going bankrupt. A variant of the above scenario, with both the participant and the casino having infinite resources, was introduced by the Swiss mathematician Nicolaus Bernoulli in 1713 in a letter to his friend and collaborator Pierre Rémond de Montmort. Here the participant places successive bets of $2, $4, $8, $16 and so on until the fair coin shows heads, at which point the game ends. (Bernoulli used ducats in his letter, but we will continue to use dollars in our discussion for notational convenience.) The participant pays nothing for lost bets; only an entry fee before the game starts. The question then is: "What is a fair entry fee for this game?"

Note that the probability that the game ends after n bets is $(1/2)^n$ and the amount won if the game ends after n bets is 2^n. Thus the expectation, or the probability-weighted sum over all values of n, is infinite. But surely nobody will pay an infinite entry fee where there is only a 1.8% chance of winning more than sixty-four dollars! This is known as the St. Petersburg paradox, after the *Journal of the Imperial Academy of Sciences, St. Petersburg* where Daniel Bernoulli (Nicolaus Bernoulli's cousin) published a paper in 1738, explaining this apparent inconsistency using the *law of diminishing marginal utility*. He suggested the use of a logarithmic utility function associated with wealth, so that the utility increases by 1 unit if the current wealth is doubled and decreases by 1 unit if the current wealth is halved.

The entry fee would be fair if it leaves the expected utility of the player's wealth after the game the same as the utility of his wealth before the game. Thus, for a person of wealth w, the fair value of the entry fee x will satisfy

$$\sum_{n=1}^{\infty} \log_2(w + 2^n - x) = \log_2(w)$$

For example, if w = $100000, then we get x = $17.56. This makes much more sense. However, German mathematician Karl Menger noted in 1934 that given any utility function, one can construct a betting scheme where the St. Petersburg paradox will reappear. Thus if successive bets are $2, $4, $16, $65536 and so on, (never mind that the next term is unrealistically large; the hundredth term in the original sequence wasn't particularly realistic either) then taking logarithms to the base 2, we get the utilities as 1, 2, 4, 8 and so on. So the paradox has reappeared even in the utility setting. Economists now require utility functions to be bounded precisely to avoid situations like the St. Petersburg paradox.

An interesting approach to this paradox was taken by William Feller, one of the pioneers of modern probability theory. Feller computed the fair entry fee to play the St. Petersburg game not just once, but k times, with each round consisting of bets with increasing stakes as described earlier. He defined the St. Petersburg game with an entry fee to be fair if the ratio of the entry fee to the gains accumulated over k rounds converged to 1 with probability 1, and proved that charging an entry fee $e_k = k \log_2(k)$ makes the game with k rounds fair. Feller's solution is purely mathematical, and does not make any economic assumptions such as the diminishing marginal utility of money.

Chapter 9: Quantum Mechanics

In 1900, German physicist Max Planck came up with a revolutionary idea that finally reconciled the discrepancies between the predicted and observed frequency spectra of electromagnetic radiation emitted by a *black body* (an idealized opaque, non-reflective object) at constant temperature. He postulated that the energy associated with electromagnetic radiation is emitted and absorbed in discrete packets, with the energy of a single packet (*quantum*) proportional to the frequency of the radiation. As the constant of proportionality, called the Planck's constant, has the extremely tiny value of 6.6×10^{-34} Joule seconds, the discreteness is noticeable only at atomic scales. Planck's insight paved the way for the development of quantum mechanics, a branch of modern physics that provides the best known explanations for many important microscopic phenomena.

Quantum mechanics soon evolved into a comprehensive theory of atomic structure, built on heavy mathematical machinery including abstract mathematical objects and operations such as Hilbert spaces and tensor products. Unlike the Rutherford-Bohr model of the atom which predicts a deterministic trajectory for an electron around the nucleus, quantum mechanics gives only the probability distribution of an electron in three-dimensional space at any given time. The probability distribution is derived from the complex-valued wave function (or quantum state) of the electron, denoted Ψ, the square of whose modulus evaluated at any point gives the probability density function of the electron at that point. This probabilistic interpretation of the wave function is called *Born's law* (named after its formulator Max Born). The wave-particle duality of all matter, including electrons, had been predicted by Louis de Broglie in his Ph.D. thesis in 1924, and was experimentally verified a few years later.

Probability and non-determinism are central to the quantum mechanical worldview. In 1927, Werner Heisenberg formulated the *uncertainty principle* of quantum mechanics, which asserts the impossibility of determining the position and momentum of a particle with arbitrary precision. Analogous restrictions exist for other pairs of complementary variables like angular momentum and angular position, or energy and lifetime. This is essentially a mathematical result in the theory of Fourier transforms, but it has far-reaching consequences in the realm of theoretical physics.

The key feature of a quantum mechanical system that distinguishes it from its classical analogue is that the former does not have a definite state until a measurement is made. According to the *Copenhagen interpretation*, the most influential (and pedagogically popular) school of thought in quantum mechanics, the system exists as a superposition – a linear combination of all possible states. When the system interacts with the external world, (for example, when a measurement is made), the wave function collapses into a definite state. Albert Einstein and Max Planck, both of whom had made pioneering contributions to quantum mechanics, found it very difficult to accept this idea. Einstein's famous assertion in a letter to Max Born, *"I am convinced that God does not play dice with the universe"*, was made as a response to the quantum mechanical worldview.

In 1935, Erwin Schrödinger, who postulated the partial differential equation governing the wave function of a quantum mechanical system, introduced a brilliant thought experiment to challenge the Copenhagen interpretation. A cat is placed in a sealed box with a vial of poison, which will be administered if and only if a Geiger counter placed in the box detects the radioactive decay of a single atom. The experiment will be run for precisely T seconds, where T is chosen such that the probability of radioactive decay of the atom within T seconds equals 1/2. Then the cat is equally likely to be dead or alive, and before the box is opened, the cat is in a superposition of these two states. Clearly, once the box is opened, we know whether the cat is alive or dead, but what is the state of the cat after the experiment has concluded and before we open the box? Is it half-alive and half-dead?

Schrödinger certainly intended this thought experiment to demonstrate the absurdity of the Copenhagen interpretation, but it was not as damaging as it appears to be. Measurement by a conscious observer is only one way that a quantum system can interact with the external world. The quantum mechanical system involving the cat may have had an earlier interaction if the particle decayed, and the wave function will collapse to the dead cat state. If the particle did not decay, then of course the cat is alive and that is exactly how it will be found when the box is opened.

Yet not everyone was convinced by this line of reasoning. An alternative school of thought called the *many worlds interpretation*, where there is no wave function collapse at all, was introduced by Hugh Everett in 1957 and continues to be championed by leading physicists of the present day like David Deutsch. As a quantum mechanical system evolves according to the probability distribution dictated by the wave function, all possible states that can occur do occur, in different worlds. We only belong to one of them at any given time, but our clones inhabit other worlds, and in some sense they are all us. There is no interaction between any of these worlds, so we don't know what is going on there, and they don't know what is going on here. The cat is dead in some worlds, alive in others. This is essentially a frequentist interpretation of Born's law.

There is also a Bayesian interpretation of quantum mechanics, which is known, not surprisingly, as *quantum Bayesianism.* Before opening the box, we have a certain belief about the probability of the cat being alive, and after opening it, we revise our belief. A quantum state is not objective reality, it is only the *idea of reality* in the mind of the participating agent, revised in the light of observations. Different agents have different degrees of belief, and different quantum states. What a postmodern approach to post-Einsteinian physics in a post-truth world!

Chapter 10: The Monty Hall Problem

Reasoning in probability and statistics is fraught with traps that occasionally imperil even experts in the field. We still carry the evolutionary baggage of our hunter-gatherer ancestors, and are simply not wired to fully grasp the subtleties of probabilistic reasoning and statistical inference. Many paradoxes in probability are similar to optical illusions – we can measure, but we cannot unsee. It is one thing to compute and convince ourselves, but quite another to understand and be at peace.

An infamous example of a probability puzzle that many people find hard to grasp, even after the solution is explained in detail, is the Monty Hall problem. Suppose you are a contestant in the famous game show *Let's Make A Deal* hosted by Monty Hall. You are shown three closed doors and are told that there is a car behind one of them and goats behind the other two. You are asked to pick a door, and the host opens one of the remaining two doors to reveal a goat. You are now given the option to either switch to the remaining closed door or stay with your original choice. You get what is behind the door you picked. (In the interest of completeness, it must be stated that in most cultures, a car is considered far more valuable than a goat.) Should you switch?

The surprising answer is that you should. The probability that you will win a car if you switch is 2/3, whereas the corresponding probability is only 1/3 if you don't switch. Most people, including many Ph.D. holders in mathematics, have difficulty accepting this, as was famously demonstrated by angry letters written to

Parade magazine columnist Marilyn vos Savant, who featured this puzzle on her column. Even the great mathematician Paul Erdős got this one wrong when he first heard it from his friend and collaborator Andrew Vázsonyi.

The reason most people believe that it should not make a difference whether you switch or not is that your first guess was as good as any, so why would you switch? The catch is that the game show host can never open a door to reveal a car, so if you choose a door with a goat behind it, then he must open the door with the other goat behind it. Thus your chances of winning if you do not switch is the same as your chances of choosing the right door with your first guess, which is 1/3. But one of the two unopened doors must have a car behind it, so your probability of winning upon switching is 2/3.

In *The Man Who Loved Only Numbers*, Paul Hoffman's delightful biography of Erdős, he quotes Vázsonyi's explanation as to how a legend like Erdős could get such an easy-looking puzzle wrong:

"Physical scientists tend to believe in the idea that probability is attached to things", said Vázsonyi. "Take a coin. You know the probability of a head is one-half. Physical scientists seem to have the idea that the probability of one-half is fused with the coin. It's a property. It's a physical thing. But say I take a coin and toss it a hundred times and each time it comes up tails. You'll say something is wrong. The coin is false. But the coin hasn't changed. It's the same coin that it was when I started to toss it. So, why did I change my mind? Because my mind has been upgraded with information. This is the Bayesian view of probability. It took me much effort to understand that probability

*is a state of mind. My hypothesis is that Erdős had this idea of probability as being attached to physical things and that's why he couldn't understand why it made sense to switch **doors**."*

This explanation makes a lot of sense, but it should not be construed as a victory for Bayesians over the frequentists. It is perfectly possible to solve the Monty Hall problem correctly using frequentist techniques. Indeed, Vázsonyi finally succeeded in convincing Erdős by running a Monte Carlo simulation, and not by Bayesian arguments. But there is something to be said about the pitfalls of assuming probabilities as immutable, and the Monty Hall problem is indeed a cautionary tale in this regard.

www.ingramcontent.com/pod-product-compliance
Lightning Source LLC
Chambersburg PA
CBHW070514220526
45467CB00002B/660